AF270245

Animals in the Ocean

Julie Murray

Abdo Kids Junior
is an Imprint of Abdo Kids
abdobooks.com

Abdo
ANIMAL HABITATS
Kids

abdobooks.com

Published by Abdo Kids, a division of ABDO, P.O. Box 398166, Minneapolis, Minnesota 55439.
Copyright © 2021 by Abdo Consulting Group, Inc. International copyrights reserved in all countries.
No part of this book may be reproduced in any form without written permission from the publisher.
Abdo Kids Junior™ is a trademark and logo of Abdo Kids.

Printed in the United States of America, North Mankato, Minnesota.

052020

092020

THIS BOOK CONTAINS
RECYCLED MATERIALS

Photo Credits: iStock, Shutterstock

Production Contributors: Teddy Borth, Jennie Forsberg, Grace Hansen

Design Contributors: Candice Keimig, Pakou Moua, Dorothy Toth

Library of Congress Control Number: 2019955581
Publisher's Cataloging-in-Publication Data

Names: Murray, Julie, author.

Title: Animals in the ocean / by Julie Murray

Description: Minneapolis, Minnesota : Abdo Kids, 2021 | Series: Animal habitats | Includes online resources and index.

Identifiers: ISBN 9781098202125 (lib. bdg.) | ISBN 9781098203108 (ebook) | ISBN 9781098203597 (Read-to-Me ebook)

Subjects: LCSH: Animals--Habitations--Juvenile literature. | Habitat (Ecology)--Juvenile literature. | Marine animals--Juvenile literature. | Ocean--Juvenile literature. | Marine ecology--Juvenile literature.

Classification: DDC 591.52--dc23

Table of Contents

Animals in the Ocean

Many animals live in
the ocean.

An octopus has eight arms.

It also has no bones.

Sea turtles can weigh

1,000 pounds (450 kg)!

Fish swim in the ocean.

Some have **bright** colors.

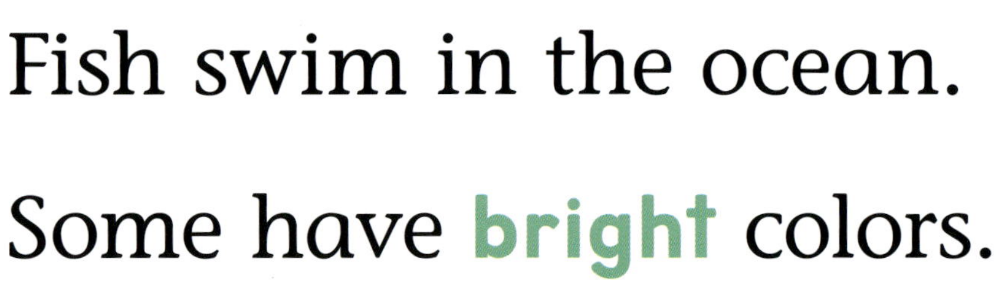

11

Sharks are fast. They can swim 40 mph (64 km/h)!

shortfin mako shark

Dolphins live in a group.

The group is called a pod.

Starfish are not fish. They are **related** to sand dollars.

sand dollars

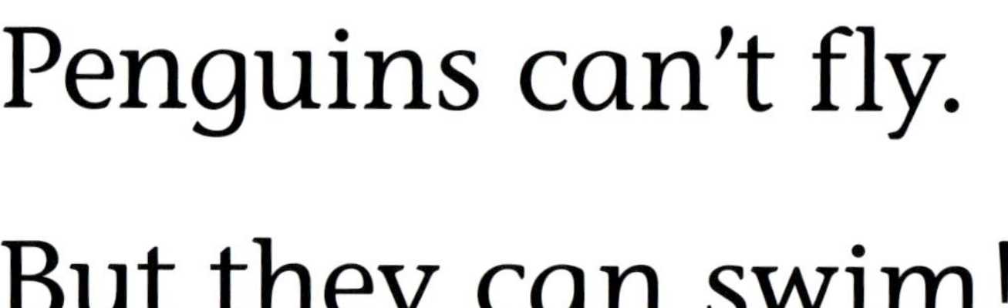

Penguins can't fly.

But they can swim!

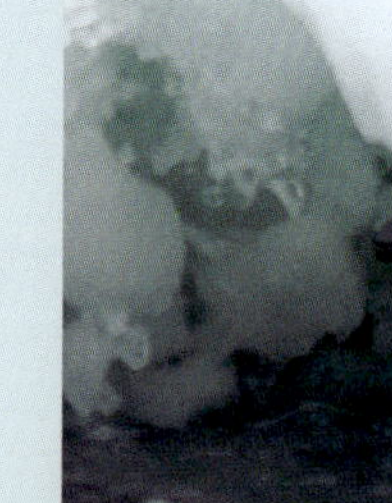

Whales are big. The blue whale is the biggest.

More Animals in the Ocean

lobster

sea horse

stingray

walrus

Glossary

bright
lively and strong.

related
belonging to the same family
or group.

Index

Visit **abdokids.com** to access crafts, games, videos, and more!